Canon EOS M50 Mark II

Camera User Guide

*Master Camera Settings, Photo & Video Basics, and
Real-Life Shooting with the Canon M50 Mark II—A
Beginner's Guide for Photography & Vlogging Success*

Georgette Howard

1

Table of Contents

11

Introduction: The Guide You Wished Came in the Box

If you've just unboxed your Canon EOS M50 Mark II or you've had it sitting in a drawer because it feels too complicated to use—you're not alone.

You probably bought this camera with excitement. Maybe it was for capturing beautiful memories, upgrading your travel photography, documenting your life on video, or even launching a YouTube channel. You wanted pro-level quality in a small, approachable package and you chose the right camera. The M50 Mark II can do all of that and more.

But then, reality hit.

You opened the Canon instruction manual and were instantly overwhelmed by hundreds of pages of technical jargon, cryptic icons, and settings that make no sense. You just want to know how to take a great picture or shoot a video without the stress. You searched YouTube for answers, only to be flooded by fast-talking tutorials that assume you already know what "aperture," "ISO," and "frame rate" mean.

If that sounds like you, this book was made for you.

No Technical Background? No Problem.

This guide speaks your language not a photographer's textbook. We've stripped away the confusion and built a beginner-first, visually rich, step-by-step companion to help you feel confident with your Canon EOS M50 Mark II from the very first shot.

Whether you're:

- A **senior learning photography for the first time**
- A **parent capturing family memories**
- A **vlogger or content creator** starting your journey

- Or just **a frustrated beginner** tired of Googling how your camera works...

You'll find clear, friendly explanations, real-world photo examples, and practical cheat sheets that show you exactly what to do without needing a tech degree.

What You'll Actually Learn (and Use)

This isn't a recycled copy of Canon's manual. It's a hands-on, human-first field guide that covers what people really want to know:

- How to take your **first great photo** without understanding photography
- How to focus, shoot, and adjust settings **visually, not theoretically**
- How to **record high-quality video** and audio for YouTube or family moments

16

- How to avoid blurry images, confusing menus, and settings that ruin your shots
- How to confidently use **modes, buttons, autofocus, lenses, and flash**
- How to **transfer, edit, and share** your photos with ease
- How to stop feeling lost and start enjoying your camera

Every chapter was written with one goal: to remove frustration and replace it with confidence.

Who This Guide Is For

📷 First-time camera users

😊 Seniors who don't want to wrestle with confusing PDFs

🎥 Beginner vloggers and creators

📱 Smartphone shooters switching to a real camera

🎁 New Canon owners who feel overwhelmed

You don't need to know anything about photography. You just need curiosity and this book. We'll meet you exactly where you are and walk with you step-by-step.

Let's Make This Camera Work for You

You've spent good money on your Canon EOS M50 Mark II. It's time you start using it with joy and purpose whether you're capturing grandkids, nature walks, family holidays, or your very first vlog.

This book was written to be your friendly guide, your personal coach, and your confidence booster—all in one.

Let's begin your photography journey together.

📷 Turn the page, pick up your camera, and let's make

magic.

Chapter 1

The Problem with the Canon Manual (and How This Guide Fixes It)

Why Most People Give Up: The Truth About That Canon Manual

Let's be honest for a moment. You bought your Canon EOS M50 Mark II expecting a simple, fun experience. The box promised brilliant photos, crystal-clear videos, and a camera that "anyone can use." But the moment you opened the instruction manual, everything changed.

It was like trying to learn a new language, one you never asked to study.

Suddenly, you're staring at 300+ pages of fine print, technical charts, symbols you've never seen before, and settings you didn't know existed. Terms like exposure compensation, digital tele-converter, or movie servo AF pop up without context or explanation. And not a single person in the manual talks to you like... well, a person.

For many beginners, this is the moment the excitement dies.

They close the manual, put the camera back in the box, and promise to "figure it out later." But later often turns into never.

You're not alone. This happens to thousands of new camera owners, especially:

- Seniors who didn't grow up with digital tech
- Parents trying to photograph their kids
- Creators switching from smartphones
- Travelers just wanting beautiful shots
- And beginners overwhelmed by "expert-level" lingo

And here's the sad truth: Canon's default PDF manual was never written for you.

It was designed by engineers. For engineers.

It assumes you already know what ISO is. That you understand what a histogram does. That you can mentally picture the difference between AF One Shot and Servo Mode... without ever showing you what the results actually look like.

That's the exact problem this guide was written to solve.

Who This Book Is For

If any of the following descriptions sound like you, you're in the right place:

✅ **Seniors**

You want to take beautiful pictures without fighting with buttons or menus. You want something easy to follow, with clear steps and big textand you want to feel smart using your camera, not confused by it.

✅ First-Time Camera Owners

You've never owned a real camera before. Maybe you upgraded from your phone. You want to know exactly how to set it up, what buttons to use, and how to make photos and videos look great without spending hours on YouTube.

✅ Beginner Photographers

You're ready to learn, but not interested in memorizing terminology. You want to shoot confidently, understand the basics, and build skills while using your camera, not reading theory for hours.

✅ Vloggers and Creators

You bought the M50 Mark II because someone said it's great for YouTube or content creation. But now you're stuck on settings, video modes, and how to make yourself look good on screen.

If any of that resonates with you this guide is going to be your lifesaver.

What You'll Learn (and What You Can Ignore)

This guide is not just about photography it's about feeling in control of your camera.

- We're not here to teach you every obscure feature Canon ever built into the M50. That's what the manual is for. This book is here to show you:

- How to take your first great photo with zero photography experience

- How to use Auto, Scene, and Manual Modes— without the confusion

- How to record beautiful video and audio, even if you've never filmed before

- How to understand your camera's buttons, settings, and lenses in plain English

- How to fix blurry shots, exposure problems, or focus issues

- How to connect your camera to your phone, transfer your photos, and share them easily

- How to stop feeling lost—and start enjoying photography again

- And here's what **you won't** waste time on:

- Overexplaining every menu setting you'll never use

- Deep technical charts with no real-world application

- Pages of fluff or theory without action

- Advice that assumes you've shot with a DSLR for 10 years

This book will give you **exactly what you need to feel confident**, at your own pace, with no pressure and no judgment.

How to Use This Book (No Tech Fluency Needed)

This book was designed to be used **with your camera in hand.**

It's not a sit-back-and-read kind of guide. It's a **walk-with-you, show-you-how** companion. Each chapter includes:

✦ **Simple language** written like a conversation, not a manual

✦ **Step-by-step photo** guides that show you what to do, not just tell you

✦ **Quick reference** cards you can print and carry

✦ **Practice projects** to help you gain real-world skills

✦ **Tips, checklists, and diagrams** built for seniors and visual learners

You can read it cover to cover, or just flip to the part you need: setting up, shooting video, choosing settings, etc.

By the end of this book, you'll not only understand your camera, you'll trust yourself to **use it every day with joy**.

Simple Photography Terms Decoded (So You Can Skip the Jargon)

Let's get one thing clear: you don't need to be technical to be a great photographer.

But the camera world is full of fancy terms that confuse people right from the start. That's why we'll break every term down as we go.

But here are just a few examples of what we'll simplify:

- **ISO** = how sensitive your camera is to light

- **Shutter Speed** = how fast the camera takes a photo

- **Aperture (f-stop)** = how much of the scene is in focus

- **Exposure** = how bright or dark your image is

- **Autofocus** = how the camera figures out what to focus on

- **4K Video** = ultra-sharp video recording (yes, it's worth it)

- **Scene Modes** = camera presets like portrait, landscape, food, etc.

- **White Balance** = adjusting for warm or cool lighting

You'll learn what each of these means when you need it, not all at once.

And we'll always explain it in plain English with pictures to match.

CHAPTER 2

Unboxing to First Shot—Complete Setup Made

Simple

What's in the Box—Each Part

Explained Simply

Congratulations!you've got your Canon EOS M50 Mark II! Before we dive into setup, let's take a clear look at what's inside the box. You'll likely find the following:

1. Canon EOS M50 Mark II Camera Body – This is the heart of the camera, where the screen, dials, and internal sensor live.

2. EF-M 15–45mm Kit Lens – A great starter lens that lets you zoom in and out for portraits, landscapes, or everyday scenes.

3. Battery Pack (LP-E12) – Rechargeable battery that powers your camera.

4. Battery Charger (LC-E12) – Plugs into the wall to charge your battery.

5. Camera Strap – Keeps your camera safe around your neck or shoulder.

6. Lens Cap – Protects the front of the lens.

7. Body Cap – Covers the camera sensor when no lens is attached.

8. User Manual (Printed or CD) – (You can ignore this— your real guide is in your hands now!)

TIP: If something's missing, contact the seller right away. It's rare, but not unheard of.

Battery

Lens

Charger

Camera
EOS M50

Memory
Card

Strap

Installing the Battery, Memory Card, and Strap — Step by Step

Let's get your camera powered up and ready.

Step 1: Charge the Battery

- Take the **battery pack (LP-E12)** and insert it into the **battery charger (LC-E12).**

- Plug the charger into a wall outlet.

- The indicator light turns **orange while charging,** and **green when fully charged** (about 2 hours).

Step 2: Insert the Battery

- Flip the camera upside down.

- Slide open the **battery compartment door** (labeled underneath).

- Insert the battery, **metal contacts first**, with the Canon logo facing outward.

- Push it until it clicks, then close the compartment.

Step 3: Insert a Memory Card

- Next to the battery slot is the SD card slot.

- Insert an **SDHC or SDXC memory card** (Class 10 or UHS-I recommended for video).

- The label should face the back of the camera.

- Push until it clicks. To remove it later, gently press and release.

Step 4: Attach the Neck Strap (Optional but Helpful)

- Slide the strap ends through the small metal loops on each side of the camera.
- Pull through the plastic buckles to secure.
- Adjust length to fit your shoulder or neck comfortably.

SAFETY TIP: Always use the strap when walking or standing with your camera. Accidents happen quickly.

Attaching the Lens Correctly

It's time to make your camera whole.

Step 1: Remove the Body Cap

- Hold the camera with the front facing you.

- Twist off the body cap by turning it counterclockwise.
- Set it aside in case you ever store the camera without a lens.

Step 2: Remove the Rear Lens Cap

- Take the 15–45mm kit lens and twist off the rear lens cap the same way.

Step 3: Align the White Dots

- Look for a small white dot on the camera body and a matching white dot on the lens.
- Line these two dots up.

Step 4: Twist to Attach

- Gently insert the lens into the mount.
- Twist it **clockwise until you hear a click**. That's the locking mechanism engaging.

Note: Do not force or overtighten. It should feel smooth.

Setting Language, Time, and Initial Preferences

The camera will prompt you to set up the basics on your first startup.

Step-by-Step:

- **Turn on the camera** using the **power switch** (rotate to ON).

- Use the **touchscreen** or arrow buttons to select your **language** (e.g., English).

- Set your date and time using the on-screen fields.

- Choose your **time zone**.

- If prompted, choose **"OK"** to confirm or proceed.

- That's it—you've just completed your first setup!

TIP: The time and date settings are important because your photos will be organized by date. Set them accurately for smoother file management later.

First Power-On Experience: What You'll See and What to Do

When you turn the camera on for the first time, here's what to expect:

- A quick startup screen with Canon's logo.

- The touchscreen will display live view—a real-time preview from the lens.

- If the lens is collapsed, the camera will display:

"Set the lens to the shooting position"

To Extend the Lens:

- Grab the **zoom ring** on the lens and rotate slightly until it **clicks into the extended position**.

- Now the camera is ready to shoot!

📷 *Try pressing the shutter button halfway—it should focus. Press fully to take your first photo!*

Setting Up Wi-Fi & Bluetooth Easily (for Phone Sync)

Let's connect your camera to your smartphone or tablet for easy photo sharing.

You'll Need:

- The **Canon Camera Connect app** (free on iOS and Android)
- A charged camera with a memory card inserted
- Your phone with Bluetooth & Wi-Fi enabled

Step-by-Step:

1. On your camera, go to:

Menu > Function Settings (wrench icon) > Wireless Communication Settings

2. Choose **Wi-Fi/Bluetooth** connection > Add a device

3. Open the **Canon Camera Connect** app on your phone

4. The camera name should appear—tap to connect

5. Accept pairing on both devices

Once paired, you can:

- Send photos wirelessly to your phone
- Remote control the camera for selfies or vlogging
- Geotag images using your phone's location

Pro Tip: Bookmark the Canon Connect app, you'll use it a lot for easy transfers.

Checklist: Quick Start Setup Card

✓☐ Unboxing & Setup To-Do List

- Charge the battery fully

- Insert battery and memory card

- Attach the neck strap

- Remove body and lens caps

- Attach lens (align dots, twist to lock)

- Turn on the camera

- Set language, date, and time

- Extend lens to shooting position

- Take your first photo

- Install Canon Camera Connect app

- Pair your camera with your phone

You're officially ready to shoot!

CHAPTER 3

Your Camera's Body—Buttons, Ports & Screens

Demystified

Front, Back, Side: What Each Button Actually Does (And When to Use It)

If you've ever looked at your camera and thought, "What does this button do?", you're not alone. Most new users feel overwhelmed. The good news? You don't need to memorize everything. Just focus on a few essentials to start, this chapter will walk you through them visually and simply.

Let's break the Canon EOS M50 Mark II into three main views:

Front of the Camera

- **Lens Mount Ring** – Where your lens attaches. It locks in with a twist.

- **Lens Release Button** – Press this and twist the lens counterclockwise to remove it.

- **AF-assist Beam/Self-timer Lamp** – Lights up for better focus in low light or flashes before self-timed shots.

- **Grip** – The right-side handle to hold the camera securely.

Top of the Camera

- **Shutter Button** – The main button for taking pictures.

Tip: Press halfway to focus, all the way to snap the photo.

- **Main Dial** – Turn this to change settings like aperture, shutter speed, or scene modes.

- **Power Switch (ON/OFF)** – Self-explanatory, but don't forget it's there!

- **Mode Dial** – Lets you switch between Auto, Manual, Video, Portrait, etc.

- **Movie Record Button (Red Dot)** – Press this anytime to start/stop recording video.

- **Hot Shoe** – Slot for attaching an external flash or microphone.

- **Flash (Pop-Up)** – Hidden inside until you press the flash release button on the side.

Back of the Camera

- **Touchscreen LCD** – Your main screen for composing, reviewing, and navigating.

- **Menu Button** – Opens the full camera settings menu.

- **Info Button** – Cycles through what appears on your screen (grid, level, etc.)

- **Playback Button** – Lets you view your photos and videos.

- **Trash/Delete Button** – Press while viewing a photo to delete it.

- **Control Ring / Directional Pad** – Use it to scroll through menus or adjust settings.

- **Set Button (Center)** – Press this to confirm your choices.

Using the Touchscreen: Tap, Swipe, Zoom, and Navigate with Ease

One of the best things about the M50 Mark II? It works like a smartphone.

Here's what you can do with the touchscreen:

- **Tap to Focus** – Touch the screen to choose your focus point.

- **Pinch to Zoom** – In playback mode, zoom in on photos just like on your phone.

- **Swipe Left/Right** – Scroll through images or videos you've taken.

- **Tap Buttons/Icons** – You can use the screen instead of the physical buttons.

- **Drag AF Box** – Move the autofocus point anywhere in the frame.

For seniors or shaky hands: Touchscreen is easier and more accurate than tiny buttons.

Understanding the Viewfinder vs. LCD Screen

The M50 Mark II gives you two ways to see your shot:

1. LCD Screen (Live View)

- What most people use

- Big, bright, and great for framing shots or videos

- Fully articulating—flip it out for selfies or vlog setup

- Touch-sensitive for easy control

2. Electronic Viewfinder (EVF)

- Eye-level screen—great for outdoor shooting in bright light

- Shows you exactly what the image will look like

- Use when the sun makes the LCD hard to see

The camera automatically switches when you put your eye to the EVF.

You can also press the EVF/LCD button to toggle manually.

What the Icons on the Screen Mean (Visual Guide)

The screen might show a lot of symbols—but don't worry. Here are the most important icons you'll commonly see:

- ▭ Battery Icon – How much charge is left
- ⏱ Shutter Speed – Controls motion blur or freeze
- *f*/Aperture – Controls background blur
- ISO – Light sensitivity setting
- 🎥 – Video recording mode active
- 📷 – Shooting mode (Auto, Portrait, etc.)

46

- ☀□/ ☽ – White balance (Daylight, Tungsten, etc.)

- ⚙□ – Settings menu access

- ▓ – Exposure meter (helps you know if the image will be too dark or bright)

We'll give you a printable icon key in the bonus section, so you don't have to memorize them.

When to Use the Flash and How to Pop It Up

The flash is hidden by default but can be manually activated.

✅ When to Use It:

- Indoor photos in poor light

- Backlit subjects (e.g., person standing in front of a window)

- Nighttime portraits or close-up shots

✅ How to Pop It Up:

Press the flash release button on the left side of the camera (next to the lens).

It flips up automatically.

Note: In "Auto" mode, the flash may also pop up on its own if needed. You can turn it off manually via settings if you prefer no flash.

Quick Reference Card: Button & Function Overview

Top Buttons to Remember First

Button	What It Does
Shutter Button	Focus and shoot
On/Off Switch	Turn camera on
Mode Dial	Change between Auto, Portrait, Manual, Video, etc.
Red Movie Button	Record video
Playback Button	Review your photos
Set Button	Confirm selections

CHAPTER 4

Understanding the Modes—From Full Auto to

Manual (Gently Introduced)

The Mode Dial: What Each Icon

Means in Plain English

Let's start with the most important dial on your camera—the **mode dial**, located on the top-right side.

It controls **how your camera behaves when you take a photo or record video**.

Instead of overwhelming you with all the details at once, we'll break down the **core modes**, what the icons mean, and when to use them in real life.

On Your Mode Dial, You'll See Icons Like:

- **A+** – Fully Automatic (Scene Intelligent Auto)

- **CA** – Creative Assist Mode

- **SCN** – Scene Modes (Portrait, Food, Night, etc.)

- **Tv** – Shutter Priority Mode

- **Av** – Aperture Priority Mode

- **M** – Manual Mode

- **Movie Camera Icon** – Movie Mode

- **Two Squares Icon** – Special Effects & Filters

- **C1/C2** – Custom Modes (for advanced use)

We'll focus on the **most useful and beginner-friendly** modes first.

You don't need to master all of these now. We'll build confidence one step at a time.

Auto Mode vs. Creative Assist vs.

Scene Modes

Let's look at the beginner modes that give you great results without needing to understand everything:

A+ Mode – Scene Intelligent Auto (Green Box Icon)

Best for: total beginners, first-time users

- The camera takes full control: focus, brightness, color, flash, and everything else.
- It's like using a smartphone camera, but with better quality.
- You just point and shoot.

📷 *Perfect for: outdoor photos, family events, pets, simple vlogging*

CA – Creative Assist Mode

Best for: new users who want control without the confusion

- Looks like Auto, but lets you adjust background blur, brightness, color tone—using words, not jargon.

- Example: Instead of "Aperture f/1.8," you'll see "Background Blur: High."

- You see live previews of changes before taking the shot.

📷 *Perfect for: portraits, food photos, testing creative looks*

SCN – Scene Mode

Best for: choosing a preset based on what you're shooting

When you turn to SCN and press SET, you can choose from options like:

- Portrait
- Landscape
- Sports
- Food

- Night Scene

- Handheld Night Scene

- Silent Shooting

- HDR Backlight Control

Each of these modes automatically adjusts the camera for the scene type.

📷 *Perfect for: beginners who want pro results with zero stress*

When to Try Aperture or Shutter Priority (Without Fear)

Now let's ease into the semi-automatic modes. These give you some control while the camera still helps out.

Av – Aperture Priority Mode

You control: depth of field (how blurry or sharp the background is)

Camera controls: the shutter speed

📷 *Use when:* You want portraits with soft backgrounds or landscapes where everything is sharp.

- Turn the main dial to make the background blurrier or sharper.
- Lower numbers (like f/2.8) = blurrier background.
- Higher numbers (like f/11) = sharper image front to back.

Tv – Shutter Priority Mode

You control: how long the shutter stays open

Camera controls: the aperture

📷 *Use when:* You want to freeze motion (sports, kids, birds) or create motion blur (waterfalls, light trails)

- Fast shutter (1/1000 sec) = freeze fast action
- Slow shutter (1/10 sec) = capture movement blur

Tip: Use a tripod for slow shutter speeds to avoid blur.

Manual Mode for the Brave (With Training Wheels)

M – Manual Mode

This is where you control everything: **aperture, shutter speed,** and **ISO**.

But don't worry we'll guide you in small, simple steps.

Start with:

- **ISO**: Set to Auto (let the camera pick it)
- **Aperture**: Use a mid-range number like f/5.6
- **Shutter Speed**: Start at 1/125 for daylight shots

Use the **exposure meter** on screen (a horizontal scale) to balance your brightness.

Just aim to keep the indicator near the center.

Tip: Practice manual mode at home with still objects in daylight.

📷 *Use when*: You want full creative control (portraits, night shots, advanced photography)

Special Modes for Vlogging, Portraits, and Slow Motion

Movie Mode

Switch to the Movie Camera Icon to record high-quality video.

Use the **flip screen** to record yourself

Press the **red Movie** button to start/stop

Frame rate and resolution can be changed in the **menu (FHD, 4K, etc.)**

📷 *Perfect for:* vlogging, tutorials, family videos

Portrait Mode (SCN > Portrait)

Softens skin tones, blurs background, and uses flattering light settings.

📷 *Great for:* kids, couples, close-ups with natural light

Slow Motion (High Frame Rate Video)

- Switch to **Movie Mode**
- In settings, choose **HD 120fps** for slow-motion video (played back at 30fps)
- Best in bright daylight

📷 *Use for*: sports, pets running, dancing, fun action shots

Understanding the Modes

A+ CA
Creative Assist
Fully automatic   Easy background blur, brightness control

SCN SCN
Scene Modes
Simple presets for   Simple presets for portraits, landscapes, etc.

SCN SCN
Scene Modes
Simple presets   You control depth of field (background blur)

M M
Manual
You control motion blur   You control both exposure and motion

▶ Movie
Video   For recording video (use flip screen for vlogging)

Special Scene Modes

▭ Special Scene Mode
Movie   For recording video (use flip screen for vlogging)

Portrait   Slow Motion
Soft skin tones, blurred background   Dramatic playback of fast action

Quick Card: Mode Reference Chart for Common Scenarios

Scenario	Recommended Mode

59

Everyday photos	Auto (A+) or CA Mode
Portraits with blurred background	CA or Av Mode
Fast-moving subjects	SCN (Sports) or Tv Mode
Night city scenes	SCN (Night) or Manual
Food photography	SCN (Food) or CA
Vlogging	Movie Mode (with face tracking on)
Family indoors	Auto with flash off or SCN (Handheld Night)
Selfies	Movie Mode + Flip Screen
Long exposures (light trails)	Manual with tripod

You're now equipped to start choosing your shooting

modes with intention—not guessing.

CHAPTER 5

Taking Your First Great Photo — Step-by-Step

Walkthrough

So far, we've learned how to set up your Canon M50 Mark II, explored the buttons, understood the modes and now it's time to take your very first intentional, beautiful photo.

And don't worry you won't need a tripod, photography degree, or perfect lighting. Just your camera, this guide, and a moment of curiosity.

How to Focus: Tap, Button, Face Tracking & Eye Detection

Getting a photo in **focus** is the single most important skill in photography. Luckily, your M50 Mark II makes it easy if you know what to expect.

You have **four reliable ways** to focus:

1. Half-Press the Shutter Button

- Gently press the shutter button halfway down.
- You'll hear a beep and see a green box around what's in focus.
- Press all the way down to take the shot.

2. Tap-to-Focus on the Touchscreen

- Works like a smartphone!
- Just tap where you want the camera to focus—like your child's eyes or a flower.

3. Face Tracking Autofocus

- Perfect for portraits, kids, or vlogging.
- Your camera will detect and track faces in real time.
- Enabled automatically in most Auto and CA modes.

4. Eye Detection Autofocus (Auto or Portrait Mode)

- Tracks the eye for razor-sharp portraits.

- Especially useful when the subject is moving or turning.

Tip: Make sure there's enough light for fast focus—dim rooms slow it down.

Composition Basics: Rule of Thirds, Framing People or Pets

A photo isn't just about clarity it's about **how things are arranged** in the frame. That's called **composition.**

The simplest composition rule? **The Rule of Thirds**.

Rule of Thirds:

- Imagine your screen divided into 9 equal squares (like a tic-tac-toe board).

- Place your subject on the lines or **where they intersect**, not directly in the center.

- Creates a more dynamic and natural-looking photo.

Your M50 can show gridlines to help. Enable them in Menu > Display Settings.

Framing Tips:

- **People:** Focus on the eyes. Don't cut off heads or limbs awkwardly.
- **Pets:** Get down to their level, don't shoot from above.
- **Outdoors:** Use trees, windows, or doorways to frame your subject.

Getting Sharp Shots (Without a Tripod)

You can absolutely take sharp, pro-looking photos handheld. Just follow these guidelines:

Tips for Tack-Sharp Images:

- **Hold the camera close to your body**, elbows in.

- **Use two hands**—right on the grip, left supporting the lens.

- **Shutter speed:** Stay at 1/100 or faster in daylight. (Auto usually handles this well.)

- **ISO:** Keep at Auto unless you're in Manual mode.

- **Lighting:** Brighter = sharper.

Bonus: Avoid zooming in too much with the kit lens—it's sharpest between 18–35mm.

Using the Touchscreen to Shoot Like a Smartphone

If you're coming from a smartphone, the touchscreen on your Canon will feel familiar and it can be used to shoot, focus, zoom, and browse:

Touch Features You'll Love:

- **Tap to focus**: Touch the screen to focus on any area

- **Touch shutter**: Enable in menu to shoot just by tapping the screen

- **Pinch to zoom**: Zoom in on playback

- **Swipe to browse**: Flick through your photo gallery

Enable Touch Shutter: Menu > Touch Shutter > Enable

Avoiding Blurry Photos — Troubleshooting in Real Time

Blurry shots are a beginner's #1 frustration. Let's solve it.

Common Causes of Blur + Solutions:

Problem	Likely Cause	Fix
Subject blur	Movement of person/pet	Use faster shutter (Tv mode: 1/250

		or faster)
Camera shake	Unsteady hands	Tuck elbows, use grip, increase ISO
Out of focus	Focused on background	Tap to focus on subject's eyes or face
Low light	Slower shutter speed	Use flash or shoot in better light
Autofocus hunting	Low contrast scene	Focus on edges, patterns, or faces

Use Playback > Magnify to zoom in and check sharpness after each photo.

Build Confidence with This Rule:

Every single sharp photo starts with **focus**, **light**, and **steady hands**.

And even the pros miss shots. What matters is practice and progress.

Taking Your First Great Photo—
Step-by-Step Walkthrough

How to Focus

HALF-PRESS SHUTTER BUTTON | TAP ON SCREEN | FACE TRACKING | EYE DETECTION

Composition Basics — BAD | GOOD

Framing People or Pets — BAD | GOOD

Hold Steady for Sharp Shots
1/100 sec or faster
No tripod needed

Troubleshooting Blurry Photos
BLURRY | SHARP
CAUSE: Motion or faster Shroc | TIP: Raise ISO for faster shutter speed

Exercise: First Photo Assignment with Prompts

Let's get hands-on. Choose one from each category below and take a photo using the mode of your choice (Auto, CA, Portrait, etc.):

Photo #1: Indoor Subject

- **Ideas:** Cup of tea by the window, your pet on a chair, family member
- **Focus:** Tap on the subject's eye
- **Lighting:** Use natural light from a window

Photo #2: Outdoor Subject

- **Ideas:** Tree, flower, street scene, architecture
- **Focus:** Use Rule of Thirds to place subject
- **Bonus:** Try Portrait or Landscape mode

Photo #3: Motion

- **Ideas:** Pouring water, child moving, swinging door

- **Mode:** Tv (Shutter Priority), try 1/500 shutter speed

- **Goal:** Freeze the action

You did it. You just practiced core skills of real-world photography: focus, light, and intention.

CHAPTER 6

Recording Video & Vlogging the Easy Way

Video is one of the Canon M50 Mark II's strongest features and one of the biggest reasons people buy this camera.

Whether you're making travel vlogs, family videos, YouTube tutorials, or just recording everyday memories, this chapter will walk you through how to set up your camera for video, what settings to use, and how to fix common issues like blurry focus or choppy clips.

Switching to Movie Mode—Which Settings Matter

To shoot video the right way, you need to be in Movie Mode not just press the red record button in Photo Mode.

How to Switch:

- Turn the **Mode Dial** to the movie camera icon 📹.

- Your LCD screen will now display a horizontal video layout.

- The red movie record button on top now works with full video controls.

In Photo Mode, video is restricted. Always switch to Movie Mode for best results.

Frame Rate, Resolution & Autofocus for Video

Let's decode the key video terms and help you choose the best settings without the jargon.

RESOLUTION

This is how sharp your video looks.

Setting	What It Means	Best Use
FHD (Full HD) 1920×1080	Standard HD quality	Everyday video, YouTube
4K (3840×2160)	Ultra sharp, more detail	Advanced users, scenic shots
HD (1280×720)	Lower quality, smaller files	Not recommended unless storage is limited

Tip: Full HD is perfect for most vloggers and everyday use. 4K is cropped and disables Dual Pixel Autofocus.

FRAME RATE

How smooth your video plays (measured in "fps"—frames per second)

Frame Rate	Effect	Use Case
24 fps	Cinematic look	Film-style content
30 fps	Smooth, natural	YouTube, interviews, family clips
60 fps	Very smooth motion	Sports, fast action
120 fps (HD only)	Super slow motion	B-roll, product shots, dancing

Recommendation: Set your camera to FHD 30fps for the perfect balance of quality and smoothness.

AUTOFOCUS FOR VIDEO

Set your camera to:

- **Movie Servo AF:** ON

- **AF Method:** Face + Tracking
- **Eye Detection:** Enabled

This setup will automatically detect your face, keep it sharp, and track your eyes even while you move.

Menu path:

Menu > AF > Movie Servo AF > Enable

Menu > AF Method > Face + Tracking

External Microphone Setup (for Clean Audio)

One of the easiest ways to ruin a good video is **bad audio**—echo, background noise, or muffled sound.

The M50 Mark II has a 3.5mm microphone jack, so you can upgrade your audio without breaking the bank.

Recommended Budget Mic (Under $30):

- Rode VideoMicro

- BOYA BY-MM1

- Movo VXR10

How to Set It Up:

1. Plug mic into the **mic input jack** on the camera's left side.

2. Slide the mic onto the **hot shoe mount** on top of the camera.

3. Turn the mic on (if battery-powered).

4. In Menu, check:

Sound Rec. > Auto or Manual (Manual gives better control for experienced users)

Pro tip: Use a "deadcat" wind muff if filming outdoors.

How to Record Yourself (Flip

Screen, Tripod Tips, Remote App)

Using the Flip-Out Screen:

- Flip the LCD toward you to **see yourself as you film**.
- Enable **Face + Eye** Tracking so the camera keeps you in focus.
- Use **Touch to Focus** on your own face if needed.

Tripod & Stabilization:

- Use a **mini tripod** (like Joby Gorillapod or ULANZI MT-16)
- If handheld, turn on **Image Stabilization** in the menu:

    Menu > IS Mode > Enable (for compatible lenses)

Control the Camera from Your Phone:

1. Install the **Canon Camera Connect App** (iOS/Android)

2. Pair your camera via Wi-Fi/Bluetooth (see Chapter 2)

3. Use the app to:

    - Start/stop recording

    - Check framing from your phone

    - Adjust focus and exposure remotely

Perfect for solo filming, interviews, and remote shooting.

Troubleshooting Video Problems (Focus Jumping, Overexposure)

Problem: Focus keeps shifting between background and face

- **Fix:** Set AF to "Face + Tracking" and tap to lock focus.

79

Problem: Video looks too bright or too dark

- **Fix:** Tap the screen and adjust exposure slider before recording.

- Try Manual Exposure Mode if needed.

Problem: Grainy or fuzzy video

- **Fix:** Add more light or reduce ISO in the menu.

Problem: Sound is too low or distorted

- **Fix:** Check your mic connection. In Menu > Sound Rec. > Set to Manual and test levels (aim for -12 to -6 dB)

Problem: Camera stops recording after a few minutes

- **Fix:** Use a fast SD card (UHS-I, Class 10 or higher). Lower cards can't keep up with HD/4K writing speed.

1. SWITCH TO MOVIE MODE

2. SET BEST VIDEO SETTINGS

Best video settings
Movie rec. size ▸ 💾 1 29.97P
Servo AF   Enable
Movie Servo AF   Face+Tracking
AF method   Eye+

SET OK

EXTERNAL MICROPHONE SETUP

MIC PORT

EXTERNAL MICROPHONE

CREATOR SETUP

Checklist: Beginner Vlogging Setup Under $100

Item	Purpose	Approx. Price
Mini Tripod (Joby/Ulanzi)	Stabilize handheld shots	$15–$25

Rode VideoMicro or Boya BY-MM1	Clearer sound	$30–$45
Wind Muff ("Deadcat")	Outdoor audio clarity	$10
64GB UHS-I SD Card	Video storage	$15–$20
Phone w/ Canon Connect App	Remote camera control	Free

Total Estimated Cost: $70–$100

You're now ready to start vlogging, creating, and recording video that looks and sounds great without feeling like you need a film degree.

CHAPTER 7

The Lens Guide—Choosing, Using, and

Understanding Your Glass

Most beginners believe that the camera takes the picture.

But here's the truth: The lens is just as important—if not more.

Your lens controls:

- What you see

- How much background you get

- How close you can get to your subject

- How much light reaches your camera

- And whether your photo looks "pro" or just okay

In this chapter, we'll demystify lenses, explain the kit lens you already own, and help you avoid beginner mistakes when buying new lenses or cleaning them.

Understanding the Kit Lens (EF-M 15–45mm) in Simple Terms

The lens that came with your M50 Mark II is called the **Canon EF-M 15–45mm f/3.5–6.3 IS STM**. Sounds complicated, but let's break it down:

What It Means:

Spec	Meaning
EF-M	It fits Canon's mirrorless EF-M mount (specific to M50 series)
15–45mm	It's a zoom lens—lets you go from wide shots (15mm) to tighter shots (45mm)

f/3.5–6.3	How much light it lets in (lower = brighter background blur)
IS	Image Stabilization—reduces blur from hand shake
STM	Stepping Motor—smooth and quiet autofocus, great for video

What It's Good For:

- Everyday photos
- Vlogging
- Landscapes
- Portraits in good lighting
- Travel shots

Tip: This lens collapses for compact storage. If your

screen says "Set the lens to the shooting position," rotate it outward to activate.

Zoom vs. Prime Lenses—What's Best for You?

When shopping for other lenses, you'll see two main types:

1. Zoom Lenses

- Can zoom in and out (like your 15–45mm)
- One lens = multiple focal lengths
- More flexible for everyday or travel

Examples: 18–150mm, 55–200mm

2. Prime Lenses

- Fixed focal length (no zoom)
- Often brighter, sharper, and better in low light
- Better background blur ("bokeh")

Examples: 22mm f/2, 32mm f/1.4

Beginners often find zooms more versatile. Primes are fantastic for portraits and pro-looking shallow focus.

How to Buy Affordable Lenses (Without Getting Ripped Off)

There are a lot of lenses out there—and a lot of overpriced or incompatible options.

Here's how to shop smart:

Safe Options for the M50 Mark II:

- **EF-M Lenses** – These are made for your camera and don't need adapters.

    - Canon EF-M 22mm f/2 STM – Small, sharp, great for portraits or vlogs

    - Canon EF-M 55–200mm – Adds powerful zoom for outdoor, nature, sports

- Canon EF-M 32mm f/1.4 STM – Professional-level portraits

What to Avoid:

- EF or EF-S Lenses – These are for Canon DSLRs. You can use them with an adapter, but they're bulkier and not ideal for casual users.
- Third-party EF-M lenses – Some are great (like Sigma), but always read reviews.

Lens Maintenance, Cleaning, and Changing (Step-by-Step)

Your lens is a precision tool—but it's easy to keep clean and working like new.

Cleaning Your Lens:

- Blow off dust with a blower (like a Giottos Rocket Blower)

- Wipe with a microfiber cloth or lens tissue (never your shirt)

- Use lens cleaning solution sparingly—apply to cloth, not directly to lens

Changing Your Lens (Step-by-Step):

- Turn off your camera – prevents static attraction of dust

- **Hold the camera face down**

- Press the **lens release button** (near the lens mount)

- Twist the lens **counterclockwise** to remove

- Align the **white dots** (or red dots) on your new lens and body

- Twist **clockwise** until it clicks

Tip: Change lenses indoors or in sheltered spots to avoid dust.

What Focal Length Means Visually

Let's say your kit lens is 15–45mm. What do those numbers mean?

Focal Length = Field of View (How much you "see" in frame)

Focal Length	What It Looks Like	Best For
15mm	Very wide	Landscapes, indoor rooms
22mm	Wide, slightly natural	Vlogging, street photography
35mm	Natural view (human eye)	All-purpose
50mm+	Zoomed in	Portraits, distant

| 200mm | Strong zoom | Wildlife, sports |

Lower number = wider shot. Higher number = tighter/closer shot.

Imagine standing in the same spot:

- **15mm** captures the whole room

- **45mm** captures one person's face tightly

KIT LENS EXPLAINED

Mount Type — EF-M Mount

STM Motor
FCleiing⁹)

Variable
Aperture
f/3.5–6.3

Image Stabi
tilization
ON | OFF

Image
Stabilization
ON | OFF

ZOOM LENS vs PRIME LENS

- Variable Zoom 15-45mm
- More Versatile, Flexible
- Better For Everyday Use

- Fixed Focal Length 50mm
- Often Sharper, Brighter
- Better For Portraits, Low Light

FOCAL LENGTH COMPARISON

| 15 mm | 22 mm | 50 mm |

| 15 mm | 32 mm | 50 mm |

LENS CARE & BUYING TIPS

Lens Care Steps

- Remove Dust
- Clean Glass

Avoid

Beginner-
Friendly
EF-M 22 mm

Super Cheap
Lenses

Avoid

✓ EF-S Lenses

✗ Super Cheap
Lenses

Swan Glass   Swan Lenses

91

You're now well on your way to understanding one of the most powerful tools in photography—your lens. And more importantly, you can now make smarter decisions when upgrading or exploring new creative options.

CHAPTER 8

Menu Settings That Matter (And the Ones That Don't)

The Canon EOS M50 Mark II is loaded with features, but let's face it: the menus are overwhelming.

There are dozens of tabs, hundreds of options, and even **pages that look like something out of a science textbook**. If you've ever opened the menu, looked around, and closed it immediately—you're not alone.

But here's the truth: **you only need a small handful of settings** to take great photos and videos.

This chapter will walk you through:

- The **only settings beginners need to know**
- A **simple menu breakdown**

- How to **customize your camera** to fit your workflow

- And how to **reset it all** if things ever go sideways

Overcoming Menu Paralysis: What Beginners Need Only

Let's make this easy: your camera's menu system is organized into colored tabs at the top of the screen.

Each tab leads to a group of related settings:

Color	Tab Purpose
Red 📷	Shooting Settings (Photo/Video)
Blue ▶□	Playback Settings
Yellow 🔧	Function Settings (Wi-Fi, screen, power)

Green ☐	Custom Functions
Star ★	Your personal My Menu

You'll mostly use Red (Photo & Video), Yellow (Display & Power), and eventually Green/Star (for custom setup).

We'll walk you through each one that actually matters.

📷 Red Tab – Photo & Video Settings

These control how your camera shoots and records.

Settings to Adjust (Recommended):

- **Image Quality:** Set to JPEG Fine (L) for best beginner results

  Menu > Red tab 1 > Image quality > L

- **Movie Rec Size:**

  FHD 29.97 (1080p at 30fps)

  Menu > Red tab 2 > Movie rec size

- **Movie Servo AF: ON**

  Keeps focus during video

  Red tab 4 > Movie Servo AF > Enable

- **AF Method: Face + Tracking**

  Best for both video and photo

  Red tab 4 > AF Method

- **Eye Detection AF: Enable**

  Helps lock focus on your subject's eyes

  Red tab 4 > Eye Detection AF

Ignore RAW, Dual Pixel RAW, HDR Movie, and Frame Grabs for now.

▶□ Blue Tab – Playback Settings

Controls how you view photos/videos on screen.

Beginner Tip:

- **Highlight Alert: ON**

  Menu > Blue tab 1 > Highlight alert

  Helps you see overexposed areas in photos

- **Auto Rotate: ON**

  Keeps photos upright when viewed on screen or computer

You can skip image rating, copyright info, or print order settings unless you're publishing professionally.

🔧 Yellow Tab – Function & Display Settings

These control how the camera behaves and looks.

Settings to Adjust:

- **Beep: ON (optional)**

  Helps confirm focus; disable if you find it annoying

- **Eco Mode: ON**

  Saves battery life

97

Yellow tab 1 > Eco mode

- **LCD Brightness: 4 or 5**

  Keep the screen bright enough for daylight use

- **Display Grid: 3x3 Grid**

  Helps with Rule of Thirds composition

  Yellow tab 3 > Grid Display

- **Date/Time/Zone** – Set correctly for file sorting

  Yellow tab 2

Skip HDMI, video system (NTSC/PAL), and sensor cleaning for now.

☐ **Green Tab – Custom Functions**

Advanced tab—skip for now.

We'll revisit this later when you're ready to:

- Assign custom buttons
- Lock preferred settings

- Adjust metering or bracketing

★ "My Menu" – Your Personal Quick Settings Tab

This is a **game-changer** for beginners. Instead of navigating deep menus every time, you can **bookmark your favorite settings** here.

How to Set Up "My Menu":

1. Go to the **Star Tab**
2. Select "**Add My Menu Tab**"
3. Tap **"Register Items"**
4. Choose from:

   - Image Quality

   - Movie Rec Size

   - Eye Detection AF

   - AF Method

   - Format Card

   - ISO Speed Settings

Now these show up in one place for quick access!

Resetting the Camera If You Get Lost

If things ever feel messed up, or you changed a setting and can't figure out how to fix it, there's an easy reset.

To Reset All Settings:

- Menu > Yellow Tab 5 > **"Clear Settings"**
- Choose:
    - **Clear all camera settings**
    - **Clear all custom functions**

Your camera will return to factory default. No harm done. It won't delete your photos.

Red Tab   Blue Tab   **Yellow** Tab   **Star Tab**

Photo & Video   Playback   Display and   Shortcut
Options   Settings   Power   your favorite
settings

BEST BEGINNER SETTINGS

Image quality
• JPEG 4

Movie rec size
• FHD 29.97P

Eye detection AF
• Enabled

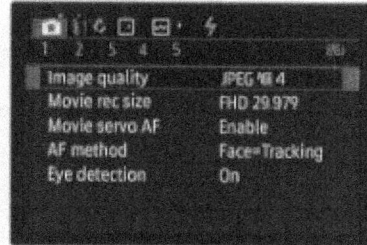

Add a Handy Shortcut Tab

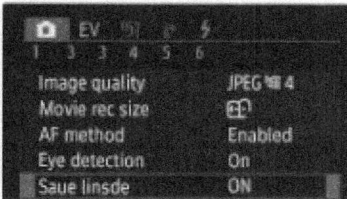

CUSTOMIZING "MY MENU"

Add a Handy Shortcut Tab

RESETTING THE CAMERA

Easy Reset to Default

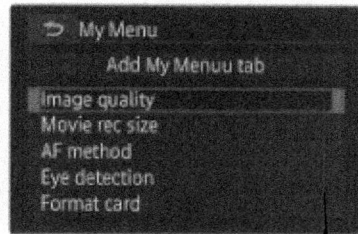

Quick Card: Settings Setup Snapshot

Essential Settings Cheat Sheet:

Function	Recommended Setting
Image Quality	JPEG Fine (L)

Video Rec Size	FHD 30fps
Movie AF Method	Face + Tracking
Eye Detection	ON
Grid Display	3x3
Beep	ON (optional)
Eco Mode	ON
Display Brightness	4 or 5
My Menu	Add 5 favorite settings

Keep this card saved on your phone for fast reference.

You now know how to navigate, customize, and control your camera's core settings—without the overload.

CHAPTER 9

Sharing, Transferring & Editing Your Photos

You've taken beautiful photos. Now what?

For many users especially beginners, the photos stay on the memory card because transferring and editing feels too complicated.

This chapter fixes that.

We'll guide you through **step-by-step** instructions to:

- Send your photos to your phone via Wi-Fi or Bluetooth
- Transfer them to your computer (Mac or Windows)
- Lightly edit for a polished look using simple tools
- Email, print, or post on social media
- And most importantly—**back up everything safely**

Sending Photos to Your Phone (Wi-Fi/Bluetooth Sync Step-by-Step)

This is the **fastest way** to share your photos with friends and family, or post them online. It works with Canon's **Camera Connect app**, available for **iOS and Android**.

What You'll Need:

- Your Canon M50 Mark II (with battery and memory card)
- A smartphone with the **Canon Camera Connect** app installed
- Wi-Fi and Bluetooth enabled on both devices

Pairing Your Camera (First Time Only)

1. **On the camera**, go to:

    Menu > Yellow Tab > Wireless Communication Settings

2. Choose:

Wi-Fi/Bluetooth connection > Add a Device

3. On your phone:

- Open the **Canon Camera Connect app**
- Select your camera model when it appears
- Accept the connection on both devices

Tip: Your phone will remember this connection for future transfers.

Sending Photos (After Pairing)

1. **Turn on the camera** and make sure Wi-Fi is enabled
2. Open the **Canon Camera Connect app**
3. Tap: "Images on Camera"
4. Browse your photos and tap to transfer
5. They'll appear in your phone's gallery—ready to post, email, or edit

Use this method to send up to 50 photos at a time—great for travel days or events.

Transferring to a Computer (Mac/Windows Guide)

Want to archive your photos or work on them with editing software?

Option 1: USB Cable (Recommended)

1. Plug your camera into your computer using the included USB cable

2. Turn on the camera

3. It will appear as a connected device

4. Open File Explorer (Windows) or Finder (Mac)

5. Drag and drop files to your desktop or desired folder

Option 2: Use a Memory Card Reader

1. Eject the SD card from your camera

2. Insert into an SD card reader (built-in or USB)

3. Browse photos and copy them to your computer

Pro Tip: Rename your folders by date or event to keep things organized

Quick Editing Tips for Stunning Results (Free or Built-In Tools)

You don't need fancy software to improve your images.

Free Tools We Recommend:

- **Snapseed (Mobile)** – Free, powerful, beginner-friendly
- **Lightroom Mobile (Free version)** – Great for brightness, color, and filters
- **Photos (Mac or Windows)** – Built-in editing for cropping, exposure, and color

What to Edit:

1. **Crop** or straighten your image
2. **Adjust brightness** or shadows

3. **Enhance color slightly** (avoid over-saturation)

4. **Sharpen** if it feels soft

5. **Add a light filter** for mood, but avoid heavy effects

Keep editing light—your original photos are already strong!

Emailing, Printing, or Posting to Social Media

Now that your photos are on your phone or computer, you can:

Email

- Open your mail app and attach photos like you would any other file
- Resize to "Medium" or "Large" for quicker uploads

Print

- Use a home printer or send to services like Shutterfly, Walgreens, or Costco

- For best results, export at full resolution

Social Media

- Instagram: Crop to square or vertical, add light filter

- Facebook: Upload directly from gallery

- YouTube (for videos): Export at 1080p, add a title and thumbnail

Tip: Add simple captions or dates to help others enjoy the context

Saving and Backing Up Your Memories

Never rely solely on one memory card or device.

Best Practices for Backup:

- Create **folders on your computer** by year/month or event
- Use an **external hard drive** (1TB or more recommended)
- Activate **cloud storage** for automatic syncing (Google Drive, iCloud, Dropbox)

Optional Workflow:

- Shoot → Transfer to phone (quick share)
- Transfer full batch to computer (organize + back up)
- Edit select favorites for posting or printing

SEND PHOTOS TO YOUR PHONE

TRANSFERRING PHOTOS & EDITING

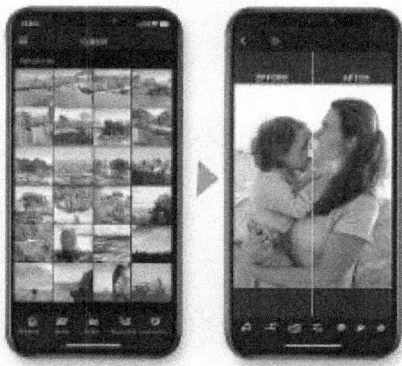

WHERE TO SHARE / PRINT

BACK UP YOUR PHOTOS

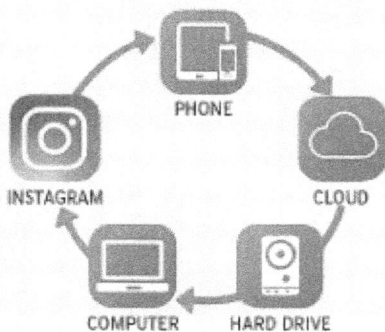

PHONE

CLOUD

INSTAGRAM

COMPUTER

HARD DRIVE

Printable: Photo Transfer & Backup Quick Guide

Your Sharing & Backup Checklist

Task	Tools	Frequency

111

Send to phone	Canon Connect App		After each outing
Transfer to PC	USB or SD Reader		Weekly or after events
Edit photos	Snapseed Photos app	/	Optional
Post online	Instagram Facebook YouTube	/ /	As desired
Backup files	External Drive + Cloud		Monthly minimum

Print this checklist and tape it to your desk or photo bag for a stress-free workflow.

You're now equipped to move your photos from camera to everywhere you want to share them confidently, cleanly, and creatively.

CHAPTER 10

Solving the Most Common Beginner Problems

Every beginner runs into roadblocks. Blurry photos. Camera won't turn on. Nothing looks right. It can be discouraging but the truth is: these issues are almost always easy to fix once you know what's really going on.

In this chapter, we'll walk through real-life beginner problems, show what causes them, and give you simple step-by-step solutions with visuals where it helps most.

Print the troubleshooting card at the end or keep it on your phone for fast fixes when you're out shooting.

Blurry Photos? Here's Why (And

How to Fix It)

This is the most common frustration for new users especially when the camera seemed to focus just fine.

Problem Types:

Type of Blur	Likely Cause	Solution
Motion blur	Subject is moving (kids, pets, etc.)	Use Shutter Priority (Tv) mode and set to 1/250 or faster
Camera shake	Your hands moved while shooting	Hold the camera tighter; increase ISO or shutter speed
Focus blur	Focus locked on background	Tap on subject's face or enable Eye Detection

Extra Tip: Use Auto ISO + Fast Shutter to let the camera adjust for sharper handheld shots.

Camera Won't Turn On or Record?

Your camera suddenly stops responding? Don't panic. Try this checklist.

Quick Fix Checklist:

- Battery fully charged? Try reinserting it.

- Power switch turned fully ON? (It can stick halfway.)

- Memory card inserted? The camera won't record without one.

- Battery/memory card doors fully closed? The camera won't power on unless both are latched shut.

- Try removing lens, then restarting. Sometimes a bad contact or incorrect mounting can freeze startup.

- Still stuck? Perform a Reset All Settings via Menu.

Keep a backup battery handy to rule out power issues.

Memory Card Full or Not Recognized?

Your screen says "Card Full" or "Card Not Recognized"? Here's what to do.

Solutions:

- Go to:

  Menu > Yellow Tab > Format Card

  (This will erase all contents—only do this after backing up photos!)

- If not recognized:

117

- Remove and reinsert the card.

- Try a different SD card (UHS-I, Class 10 or higher).

- Make sure the lock switch on the side of the card is not set to LOCK.

Avoid using the same card in multiple devices without formatting.

Can't Focus? Jumping Autofocus Explained

Autofocus issues often feel like your camera is fighting you, here's how to make peace with it.

Problem Types:

Issue	Cause	Fix
Focus hunts	Low contrast or	Switch to Face +

118

back and forth	wrong focus mode	Tracking and tap to focus
Focus locks behind subject	Background has more contrast	Tap screen to tell camera what to focus on
Camera won't focus at all	Lens too close to subject	Step back slightly or use Macro Mode (SCN)
Eye not sharp in portraits	Eye Detection OFF	Enable Eye Detection AF in AF menu tab

Good light improves autofocus dramatically.

Overexposed or Underexposed

Images? Here's the Fix

Sometimes your photo looks:

- Too bright (washed out sky, pale skin)
- Too dark (shadowy, unclear)

This is an exposure issue.

Fix Options:

1. **Tap the screen** before shooting → Adjust **exposure slider** (plus or minus)

2. **Use Exposure Compensation:**

   - Menu > Exposure Compensation > Slide left (-) to darken or right (+) to brighten

3. **Use SCN modes:** Try **Backlit HDR** or **Handheld Night Scene** for automatic exposure correction

Turn on the "Highlight Alert" to spot blown-out (overexposed) areas in your shots.

120

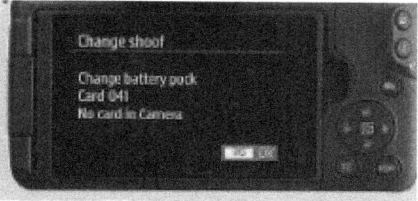

| CAUSES | Motion blur
Out of focus | CAUSES | Faster auto-on (11 250 sec or higher) |
| FIXES | ▸ Use faster shutter speed
▸ Hold camera steady
▸ Half-press shutter b focus
▸ Enable Eye Detection | FIXES | Turn on camera off-on-again
Recharge battery
Insert card
Format card |

| **OUT-OF-FOCUS** | **TOO BRIGHT OR DARK** |

| CAUSES | Wrong focus mode
Focus point shifts | CAUSES | Image is too bright
Image is too dark |

Troubleshooting Card: "If This Happens, Try This"

Problem	Fix
Photo is blurry	Use faster shutter, tap to focus, enable Eye Detection

121

Camera won't turn on	Check battery, doors, memory card, power switch
Focus keeps jumping	Switch to Face + Tracking, use touch-to-focus
Memory card error	Format card, check lock switch, try new card
Overexposed image	Adjust exposure slider or use Backlit SCN mode
Underexposed image	Increase ISO, tap to brighten, avoid shooting into light

Print and keep this card in your camera bag or photo pouch.

You've now got your own personal troubleshooting

toolbox—ready to solve any of the common issues that can trip up new camera users.

CHAPTER 11

Practice Projects for Building Confidence

By now, you've learned how to operate your Canon M50 Mark II, adjust the settings, shoot your first photos and videos, and troubleshoot common issues. But here's a truth that separates people who simply own a camera… from those who use it:

You only get good at photography by practicing.

Not with pressure. Not with perfection. But by simply picking up the camera and trying something day by day.

This chapter gives you a series of easy, feel-good photo and video projects designed to:

- Build your confidence through hands-on learning
- Reinforce what you've already learned
- Spark creativity without overwhelm

- Help you see visible progress

And you can do it all in just **7 days or 1 weekend at a time**.

7-Day Photography Challenge for Absolute Beginners

You don't need perfect lighting or expensive gear. Just commit to **1 photo per day** using the modes, tools, and techniques you've learned so far.

Here's your no-stress challenge:

✅ **Day 1: Your Favorite Object**

- Use **Auto or CA Mode**
- Focus tip: Tap on the object to focus
- Lighting: Place near a window

✅ **Day 2: A Smile**

- Use **Portrait Mode**
- Try Face Tracking + Eye Detection
- Use natural light from the side

✅ Day 3: A Meal or Cup of Tea

- Use **SCN > Food Mode**
- Shoot from directly above or 45° angle
- Focus on the main detail (center or topping)

✅ Day 4: A Shadow or Reflection

- Use **Creative Assist Mode**
- Try black & white or vivid color
- Look for puddles, windows, or mirrors

✅ Day 5: A Close-Up Texture

- Use **SCN > Close-Up or Macro**
- Texture idea: hands, fabric, leaves, wood grain

✅ Day 6: Something in Motion

- Use **Tv Mode and try 1/500 shutter speed**

- Subject ideas: pet, child, blowing curtain, water

✅ Day 7: A Self-Portrait or Vlog

- Use Movie Mode with flip screen

- Focus: Enable Eye Detection

- Bonus: Record a 30-second video introducing yourself!

Tip: Print the included "Photo Challenge Planner" and check off each day's assignment.

Indoor Photo Ideas (Family, Pets, Flowers)

You don't need to leave the house to create beautiful images.

✅ Family Member Portrait

- Use Portrait Mode or Av Mode (f/4.0 to f/5.6)

- Face a window (not directly in sunlight)

- Use Eye Detection AF

✅ Pet in Action

- Use Tv Mode, 1/250 to freeze motion

- Get low—at your pet's eye level

- Use burst mode to capture multiple frames

✅ Flower or Plant Close-Up

- Use Macro Scene Mode

- Tap-to-focus on center of flower

- Try backlighting for a glow effect

Indoor light too dim? Move closer to a window or use a table lamp with a warm bulb.

Outdoor Practice Tips (Golden Hour, Nature, Streets)

Get outside, breathe fresh air, and take your camera with you.

✅ Golden Hour Landscapes

- Time: 1 hour after sunrise or 1 hour before sunset

- Mode: SCN > Landscape or CA Mode

- Use the 3x3 grid to apply Rule of Thirds

- Focus: Tap on distant trees or horizon

✅ Nature Walk

- Subject ideas: Leaves, textures, shadows, flowers, animals

- Lens: Zoom in to 45mm for tighter detail

- Composition tip: Include foreground, middle, and background

✅ Street Shots

- Use Tv Mode for moving people or traffic (1/500 or higher)

- Keep your camera ready and shoot from the waist or shoulder

- Try black-and-white for creative flair

Avoid using flash outdoors, use natural light instead.

Self-Portrait & Video Project (Using Tripod or Remote)

You'll learn more about focus, lighting, framing, and storytelling from shooting yourself than you might think.

✅ Self-Portrait

- Use the **flip screen** and Face Tracking
- Set camera on a tripod or firm surface
- Use 10-second timer or **Canon Camera Connect app**

✅ Mini Video Intro

- Go to Movie Mode, use Face + Eye Detection
- Film a 30–60 second clip:
  - Introduce yourself
  - Share why you bought the camera

- Say what you're excited to shoot

Keep it casual and light. This builds both video and speaking confidence.

Journaling Your Progress (How to Track Growth)

Documenting your photography journey helps you:

- Notice what you're learning
- Stay motivated
- Reflect on how far you've come

✅ **Ideas for Your Journal:**

- Write down the mode or setting you used
- What worked well? What didn't?
- What did you learn from that shot?
- Rate your own photo 1–5 for fun
- Print one image per week and tape it in!

Use the printable "Challenge Planner & Journal Page" provided with this chapter.

Photography is a practice—not just a button. These small projects will do more for your confidence than any technical manual ever could.

CHAPTER 12

Recommended Accessories & Upgrades for Beginners

One of the most common questions new camera owners ask is:

"What gear should I buy next?"

Unfortunately, many people spend money on the wrong accessories, often based on internet hype or overpriced bundles that offer more confusion than clarity.

This chapter gives you a clear, no-fluff checklist of what you actually need and what you can skip so you can start small, shoot smarter, and upgrade wisely.

The Only Accessories You Actually

Need

You don't need a backpack full of gadgets to take amazing photos. Just a few well-chosen items will:

- Improve your stability

- Protect your camera

- Expand your creativity

- Make shooting more enjoyable

Here's your **starter list:**

Must-Have Accessories:

1. **Extra Battery (LP-E12 or Canon-branded equivalent)**

   One battery = 200–300 shots. You'll need a backup for travel days or video.

2. **64GB or 128GB SD Card (UHS-I, Class 10 or V30)**

   For photo & Full HD video. Look for Sandisk Extreme

or Lexar Professional.

3. **Microfiber Cleaning Cloth + Air Blower**

Keep your lens spotless. Never use shirts or tissues.

4. **Basic Camera Bag**

Protect your gear. Look for padding, weather resistance, and easy access.

Best Budget Tripods, Mic, SD Cards, Lens Hoods

Tripods:

Type	Use	Budget Option
Mini tripod	Tabletop, vlogging	Ulanzi MT-16, Joby GripTight
Light travel	Photoshoots,	Amazon Basics 50", K&F

tripod	long exposure	Concept	
GorillaPod	Wrap around objects, odd angles	Joby 3K	GorillaPod

Choose a tripod with a ball head for flexible angles.

Microphones:

Type	Best For	Budget Pick
Shotgun mic	Vlogs, interviews	Rode VideoMicro, BOYA BY-MM1
Lavalier mic	Talking directly, hidden mic	Movo LV1, Purple Panda

Plug into **the 3.5mm mic jack** on the M50 Mark II. Adds instant clarity to voice recordings.

SD Cards:

- **Minimum:** UHS-I, Class 10, V30 for Full HD video

- **Best Value:**

  - Sandisk Extreme 64GB or 128GB

  - Lexar 1066x

  - Samsung Evo Select

Avoid off-brand cards or anything below Class 10.

Lens Hoods:

- Reduce lens flare from sun or strong lights

- Protect your lens from scratches or bumps

- For **15–45mm** kit lens: Get a Canon EW-53 or compatible aftermarket hood

Inexpensive, but worth every cent.

What to Avoid as a Beginner (Don't

Waste Money Yet!)

There's a lot of gear that sounds useful—but will just sit in your drawer.

Here's what you **don't need right now:**

- **UV Filters** – Often reduce image quality and add glare

- **Giant camera bundles** – Usually filled with low-quality accessories

- **Lens converters** – Usually degrade image quality

- **Overly large backpacks** – Too bulky unless traveling with multiple lenses

- **Complicated studio lights** – Natural light works great for beginners

Stick to tools that match your current skill and needs.

Affordable Gear for Vlogging,

Travel, and Home Use

Ready to go beyond the basics?

Here's a setup you can build affordably:

Vlogging Setup

- Mini tripod (Ulanzi, Joby)

- Rode VideoMicro or BOYA mic

- Wind muff ("deadcat" cover)

- 64GB SD card

- Camera Connect App (free)

Travel Setup

- Light tripod (K&F Concept)

- Extra battery + charger

- Compact camera bag

- Lens hood for protection

- USB SD card reader

Home Setup

- Table tripod
- Ring light or soft LED panel (Neewer or UBeesize)
- Lavalier mic
- Backdrop or clean wall
- Cleaning kit

Build one setup at a time, don't try to buy everything at once.

How to Set Up a Starter Studio at Home

Want to record videos, take portraits, or shoot products indoors?

Here's a beginner-friendly home studio you can set up in a small space:

Basic Home Studio Checklist:

- Camera on tripod (eye level)

- Natural window light or LED ring light

- Neutral background (white wall, curtain, or backdrop paper)

- External mic (shotgun or lavalier)

- Camera Connect App – to monitor yourself on your phone

- Self-timer or remote shutter – for hands-free photos

Avoid shooting under ceiling lights—they cast harsh shadows. Use side light instead.

RECOMMENDED ACCESSORIES & UPGRADES FOR BEGINNERS

LENS HOOD, USB-I SD CARD, MINI TRIPOD, SPARE BATTERY, EXTERNAL MIC, CAMERA BAG, MICROFIBER CLOTH & BLOWER

Checklist: Beginner's Gear Buyer's List

Print or screenshot this to take shopping or compare online:

Item	Why You Need It	Notes

Extra battery	Longer shooting sessions	LP-E12 or Canon-approved
SD Card (64–128GB, Class 10)	Store more photos & video	Sandisk, Lexar, Samsung
Tripod (mini or full)	Stability for sharp shots & video	Ulanzi, Joby, Amazon Basics
Microphone (optional)	Clearer voice/audio	Rode, Boya, Movo
Cleaning kit	Keep lens spotless	Air blower + microfiber cloth
Lens hood	Reduces glare, protects lens	EW-53 for kit lens
Camera bag	Safety while	Small, padded,

You're now equipped to shop smart, avoid unnecessary gear, and build a functional setup that supports your real needs, whether you're shooting at home, in nature, or on the go.

Full Glossary of Beginner Camera Terms

Aperture (*f*-stop): Controls how much light enters through the lens and how blurry the background is. Lower f-number = more blur.

Autofocus (AF): The camera's automatic system for keeping subjects sharp. Includes options like Face + Tracking or Eye Detection.

Exposure: The overall brightness or darkness of an image, controlled by ISO, shutter speed, and aperture.

ISO: Adjusts the camera's sensitivity to light. Higher ISO = brighter image (but also more grain).

Shutter Speed: Determines how long the camera's shutter stays open. Fast shutter freezes motion, slow shutter creates blur.

White Balance: Adjusts the colors of your image to match different lighting (e.g., daylight, tungsten, cloudy).

Depth of Field: The area of an image that appears in focus. Wider apertures (like f/2.8) create shallow depth of field.

Metering: How your camera measures light in a scene to choose exposure settings.

JPEG vs. RAW: JPEG is processed and ready to share. RAW contains uncompressed data for advanced editing.

Burst Mode: Captures several images quickly in succession. Useful for action shots.

Face Detection / Eye Detection: Automatically finds and focuses on the subject's face or eyes.

4K / 1080p (FHD): Resolution levels for video. 4K is ultra-HD, while 1080p is standard full HD.

Image Stabilization (IS): Reduces blur caused by hand shake or minor movement.

Manual Mode (M): You control shutter speed, ISO, and aperture yourself—used for full creative control.

Camera Maintenance & Cleaning Tips

Your Canon M50 Mark II will last for years if you care for it properly. Here's how:

Lens Cleaning

- Use an air blower to remove dust

146

- Wipe gently with a microfiber cloth (never use tissue or your shirt)
- Use lens cleaning solution sparingly, applied to the cloth—not the lens

Screen & Viewfinder

- Wipe gently with a dry, clean microfiber cloth
- Avoid using liquid sprays directly on the screen

Sensor Cleaning

- Enable Auto Sensor Cleaning in menu (usually runs at power on/off)
- For manual cleaning, let a professional handle it unless you're experienced

Storage Tips

- Always use a lens cap and body cap when not shooting
- Store in a dry, cool place (use silica gel packs in your bag or case)

147

- Avoid extreme temperatures and high humidity

Final Words

You've done it. You've gone from camera confusion to capable, confident shooting. Whether you're capturing moments with family, expressing yourself through video, or just enjoying the process of learning—this is just the beginning.

Now go shoot something beautiful. 📷

Acknowledgement

To every aspiring photographer and filmmaker who dares to pick up a camera and tell a story, this book is for you. Special thanks to my family and friends for their encouragement, and to the creative community whose passion inspires me daily. Your support made this guide possible.

www.ingramcontent.com/pod-product-compliance
Lightning Source LLC
Chambersburg PA
CBHW031858200326
41597CB00012B/462